I0059364

PROTOZOA

Rebecca Woodbury, Ph.D., M.Ed.

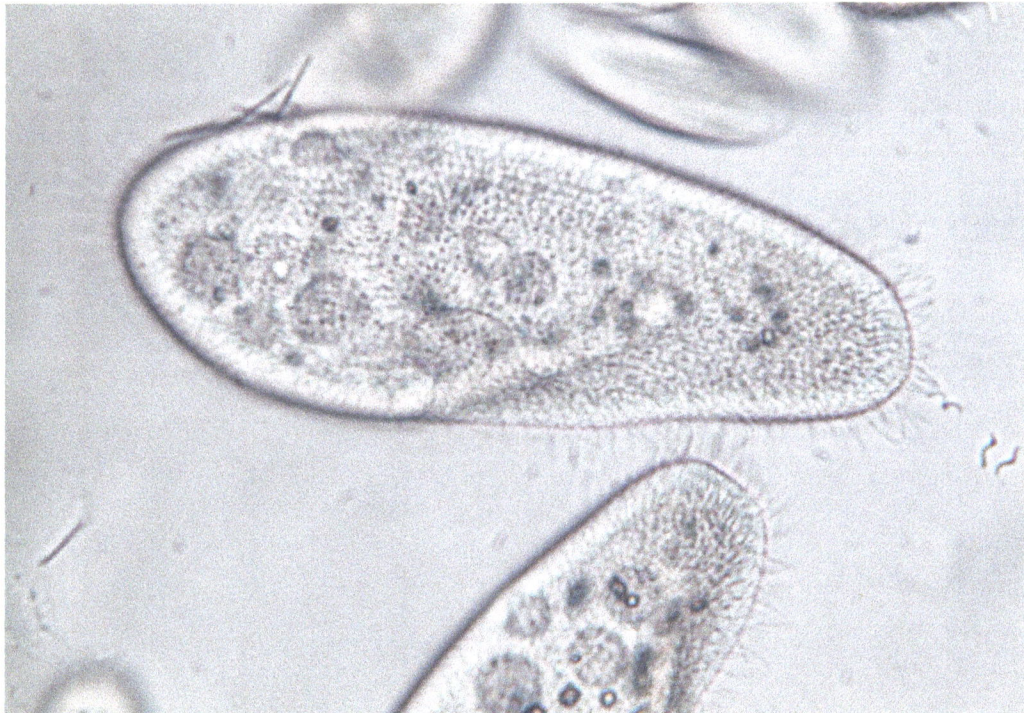

Gravitas Publications Inc.

PROTOZOA

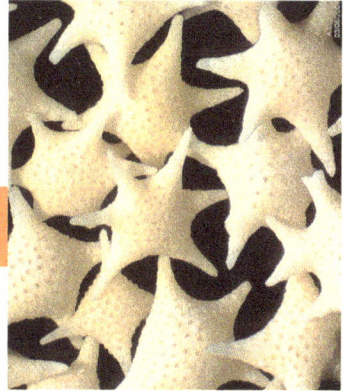

Illustrations: Janet Moneymaker

Protozoa
ISBN 978-1-950415-53-3

Published by Gravitas Publications Inc.
Imprint: Real Science-4-Kids
www.gravitaspublications.com
www.realscience4kids.com

Photo credits: Cover and Title Page: By sinhyu, AdobeStock; Above, Foramanifera, Psammophile (CC BY SA 3.0); P.7: 1. Giant kelp, Claire Fackler, CINMS/NOAA; 2. Didinium nasutum, Gregory Anitpa, San Francisco State University; 3. Dinoflagellate, Dr. John R. Dolan, Laboratoire d'Oceanographique de Villefranche; Observatoire Oceanologique de Villefrance-sur-Mer; 4. Foramanifera, Psammophile (CC BY SA 3.0); 5. Dinoflagellate, CSIRO; P.11. Paramecium, By sinhyu, AdobeStock; Amoeba, By micro_photo, AdobeStock; Euglena, By tonaquatic, AdobeStock

Protozoa are tiny creatures.

Protozoa are too small to

see with the eyes only.

I'm happy that we're not that tiny.

Yes!

Anton van Leeuwenhoek
discovered protozoa in pond water.

Leeuwenhoek looked through a
microscope to see the protozoa.

A microscope is a tool that makes
tiny things look bigger.

I want to look!

There are thousands of protozoa with different shapes and sizes.

Look what you can see with a microscope!

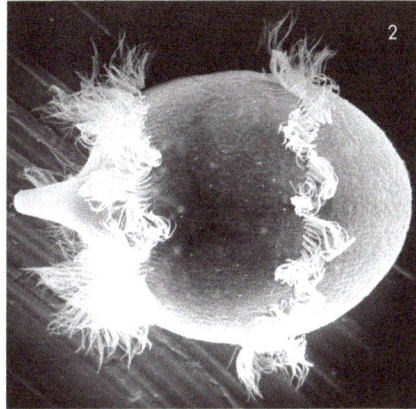

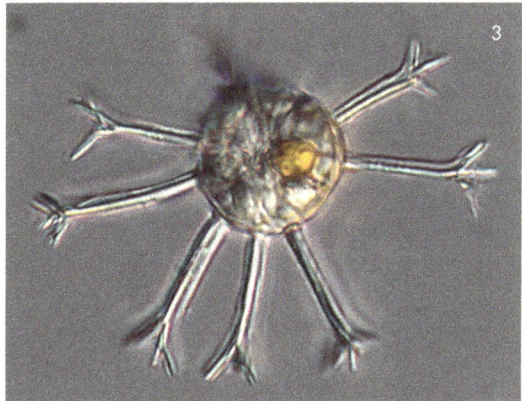

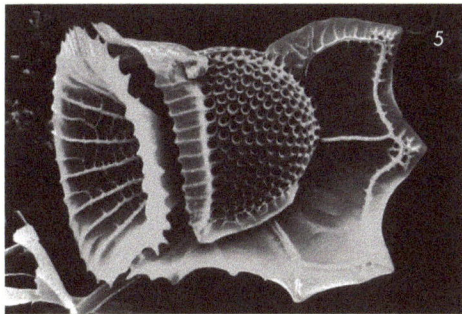

Protozoa have funny names.

Find out how to say their names.
Look in the back of this book.

This protozoan is called
a **paramecium.**

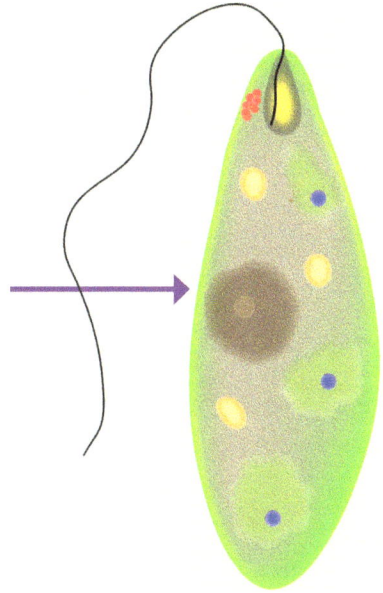

This protozoan is
called a **euglena.**

This protozoan
is called an
amoeba.

Protozoa live in water and move in different ways.

We can swim. Can you?

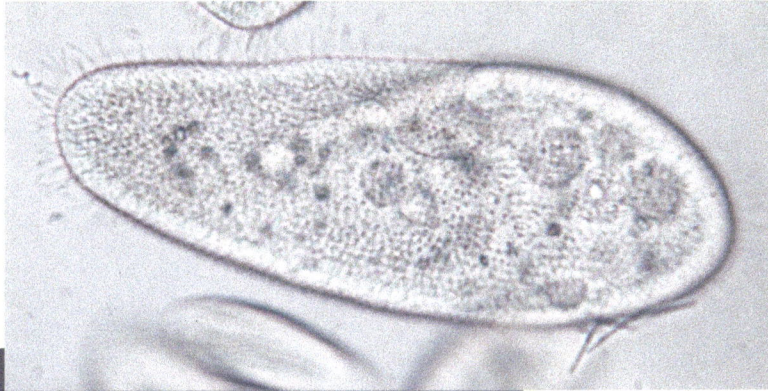

Paramecium

Amoeba

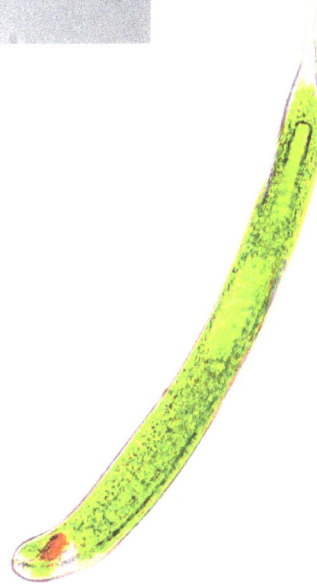

Euglena

A **euglena** swims by twirling a long tail called a **flagellum**.

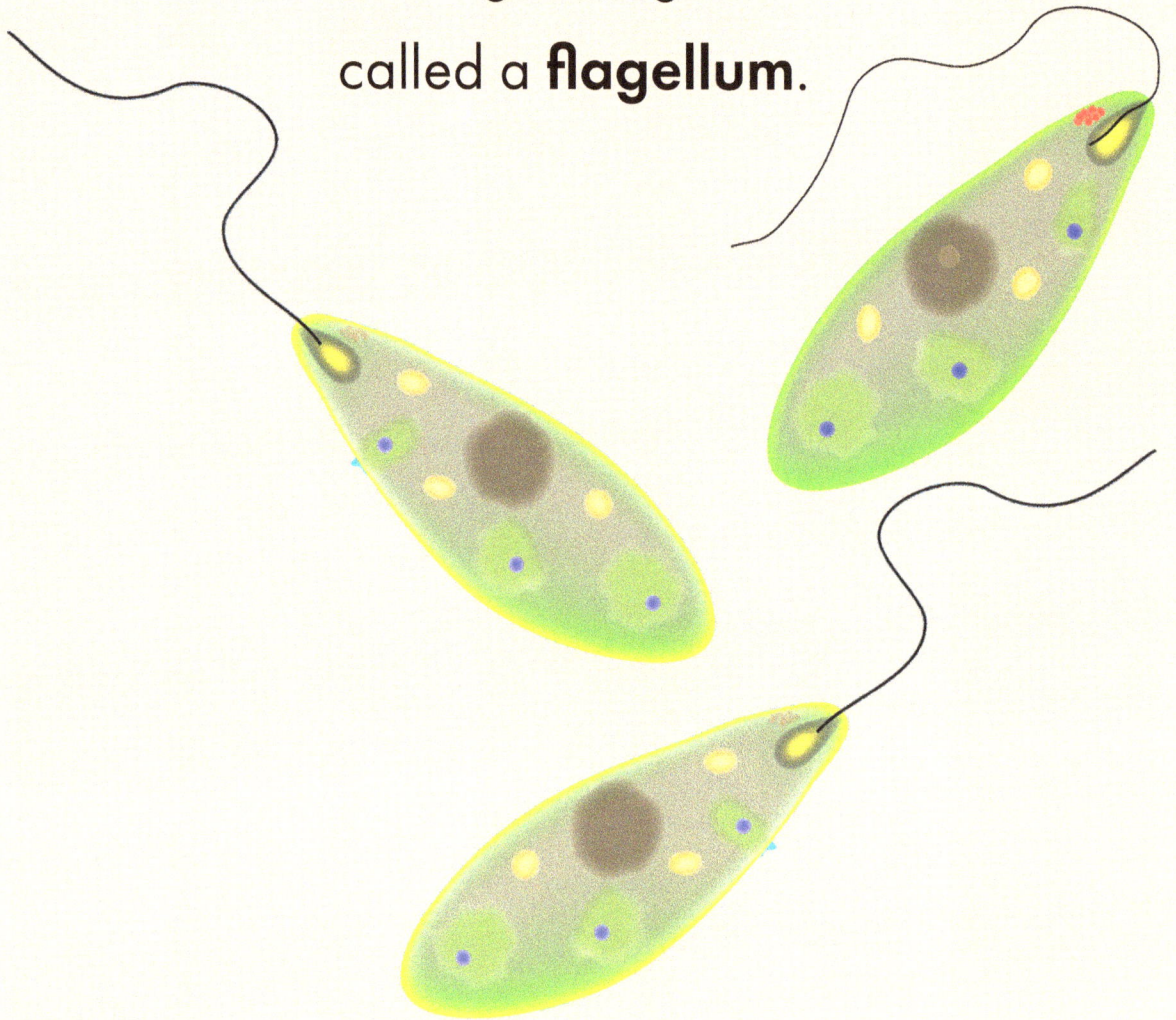

A **paramecium** swims
by wiggling tiny hairs
called **cilia.**

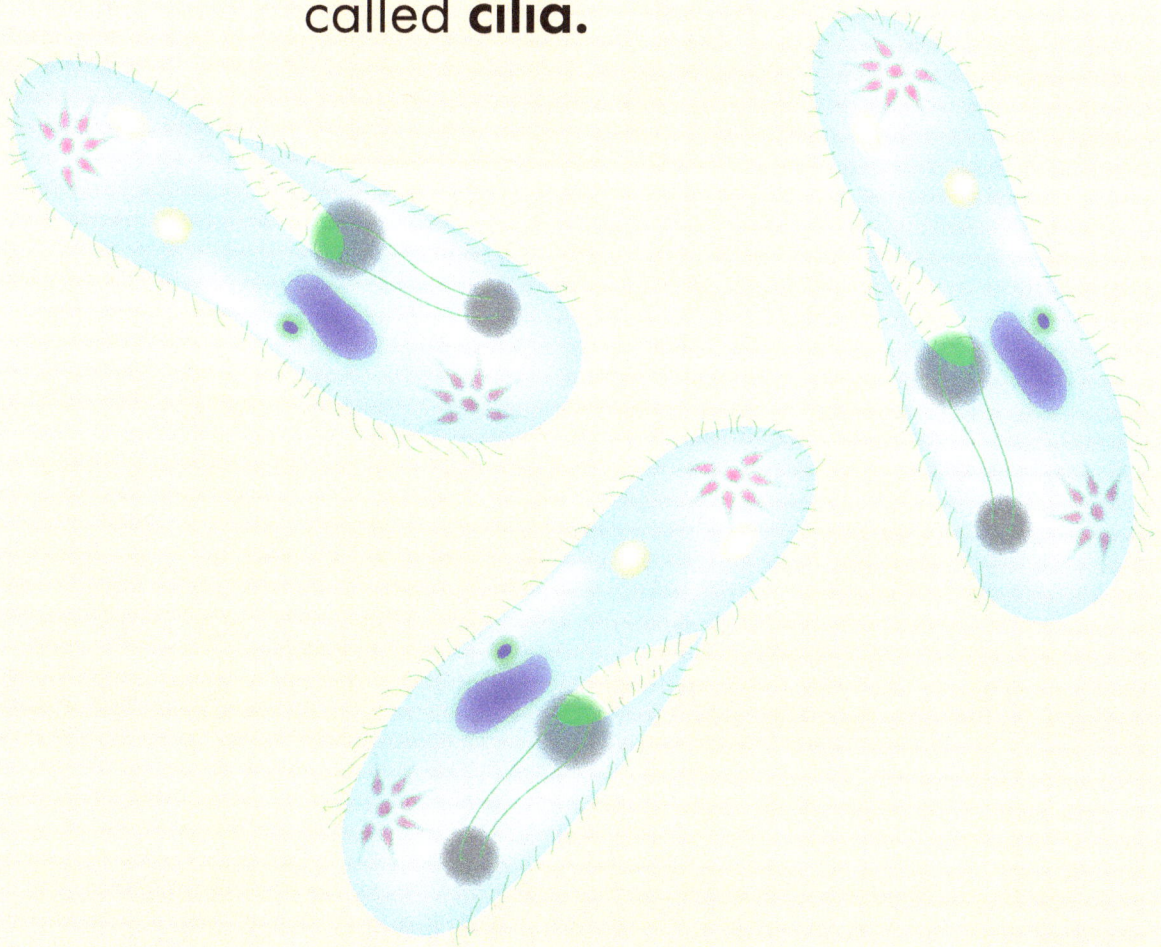

An **amoeba** crawls along a surface by using "false feet."

Are your feet real?

I think so.

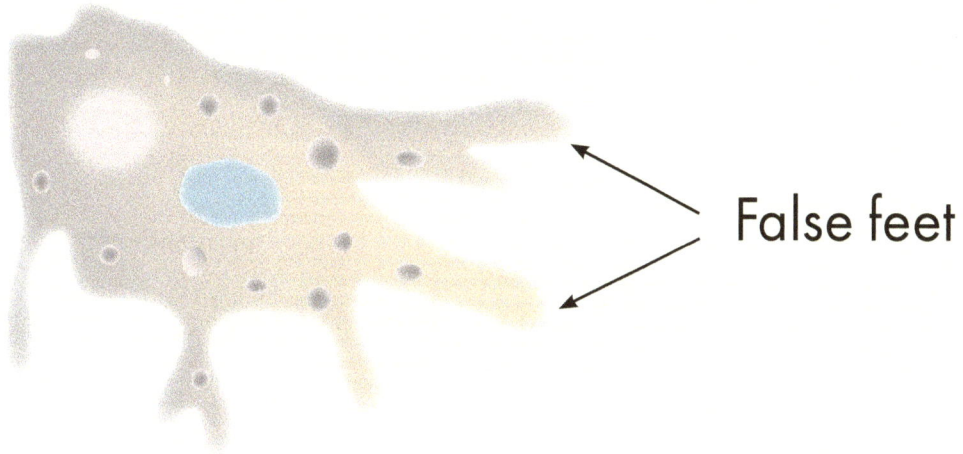

False feet

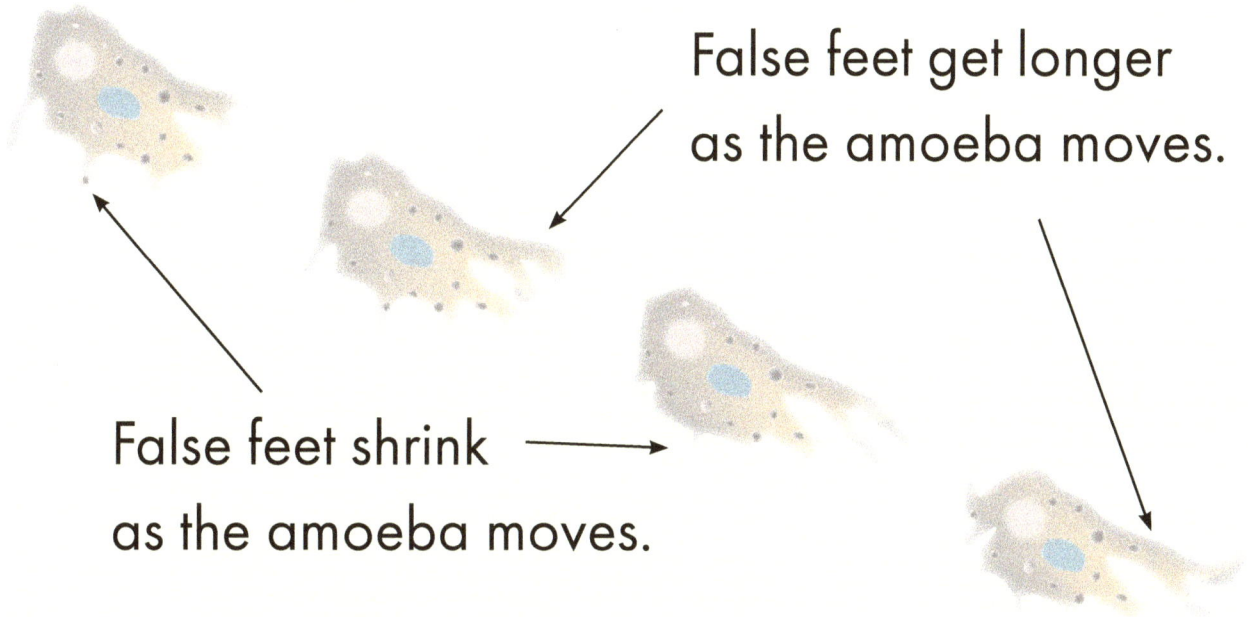

False feet get longer as the amoeba moves.

False feet shrink as the amoeba moves.

Protozoa need to eat!

Do you think they eat cheese?

Maybe.

- **Euglena** use sunlight to make their own food.

Euglena have an **eyespot** that can see light. The eyespot tells the euglena where to find sunlight.

The eyespot when it sees light.

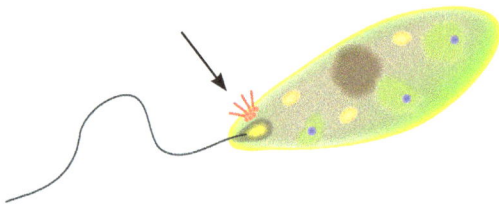

The eyespot when it does not see light.

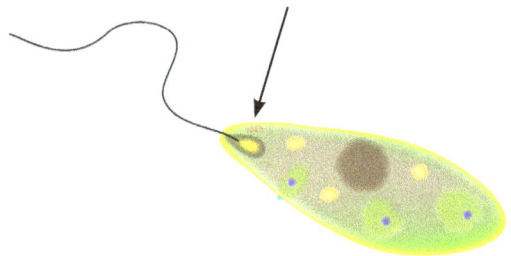

- A **paramecium** uses cilia to sweep food into its mouth.

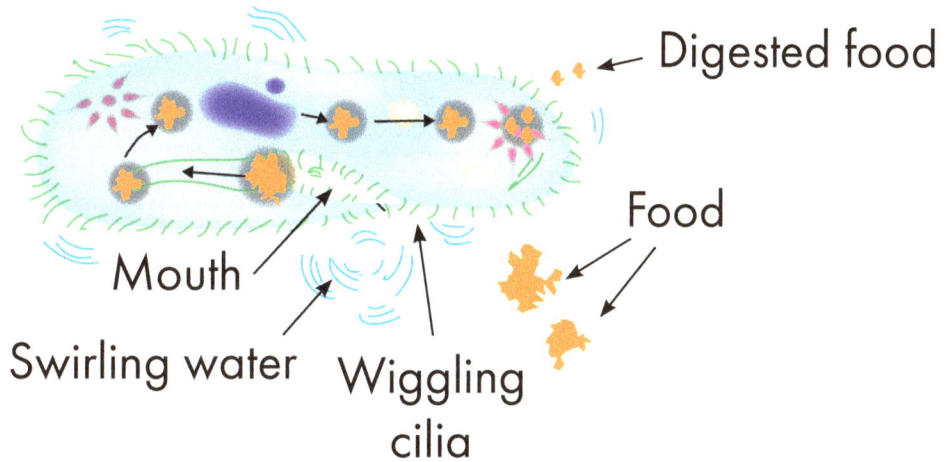

Digested food

Mouth

Food

Swirling water

Wiggling cilia

- **Amoebas** catch food with their false feet.

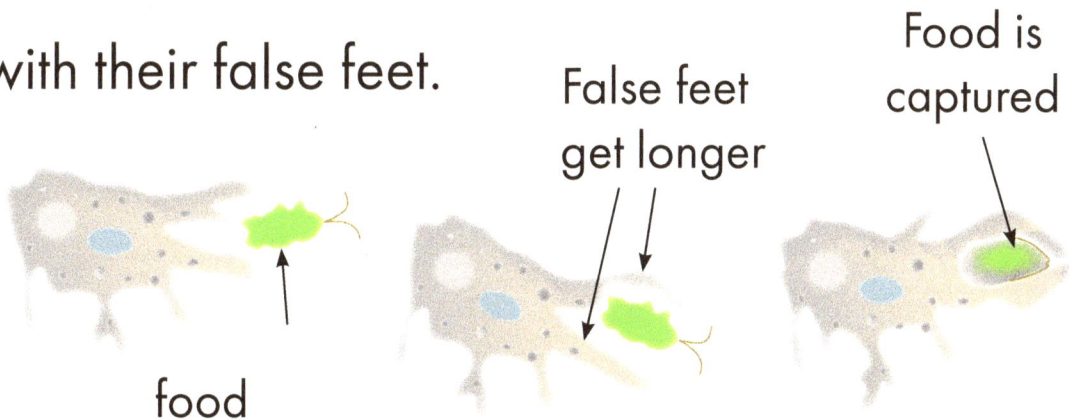

Food is captured

False feet get longer

food

- **Protozoa** live, swim, and eat even though they are tiny.

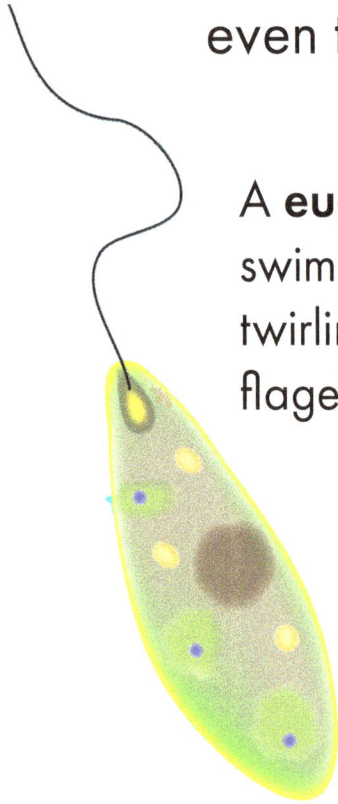

A **euglena** swims by twirling its flagellum.

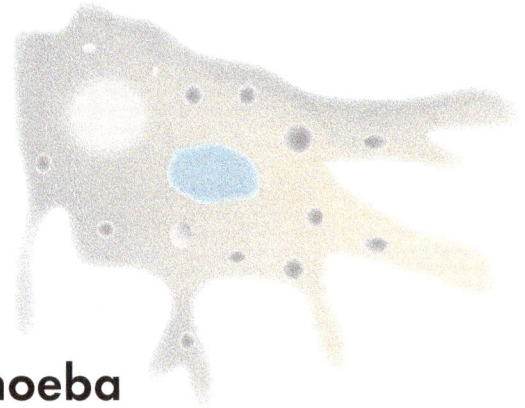

An **amoeba** crawls using false feet.

A **paramecium** swims by wiggling its cilia.

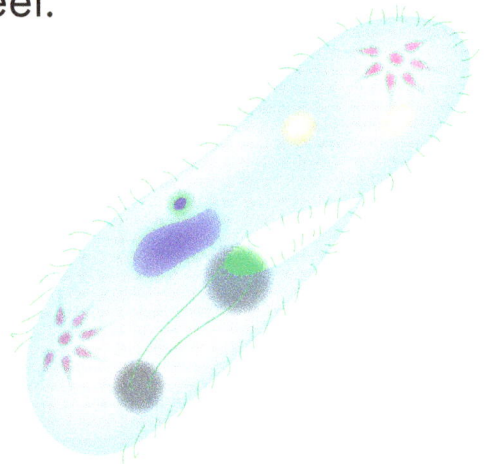

Paramecia use cilia to capture food.

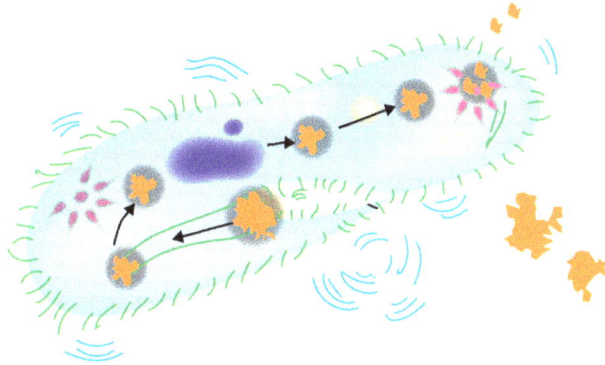

Amoebas use false feet to capture food.

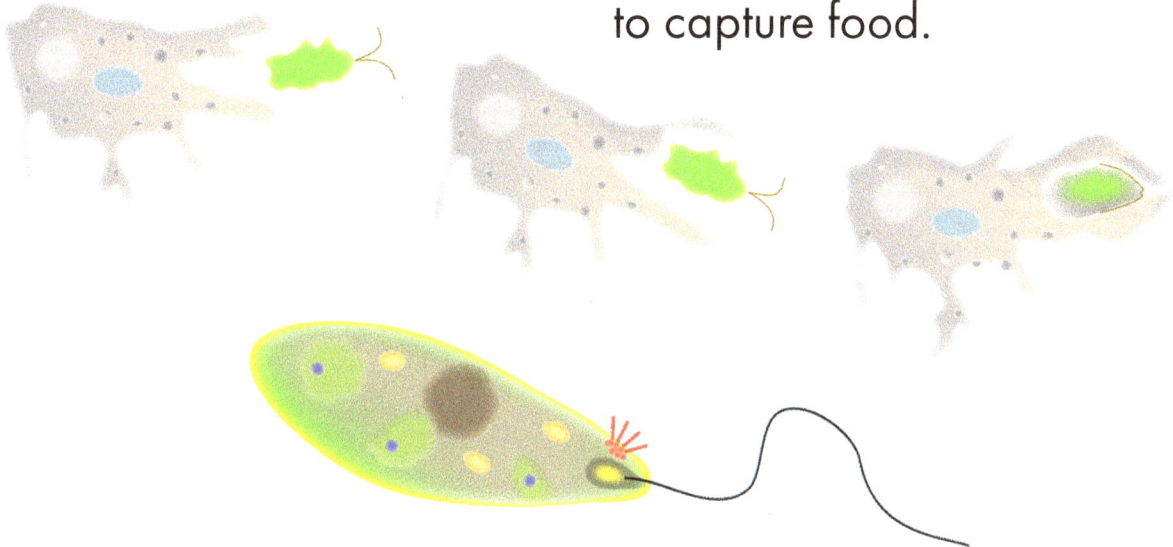

Euglena use their eyespot to find the sunlight they need for making food.

How to say science words

amoeba   (uh-MEE-buh)

Anton van Leeuwenhoek   (AN-tahn van LAY-vuhn-huhk)

cilia   (SIH-lee-uh) [plural]

cilium   (SIH-lee-uhm) [singular]

euglena   (yoo-GLEE-nuh) [singular and plural]

flagella   (fluh-JEH-luh) [plural]

flagellum   (fluh-JEH-luhm) [singular]

microscope   (MIY-kruh-skohp)

paramecia   (per-uh-MEE-shee-uh) [plural]

paramecium   (per-uh-MEE-shee-uhm) [singular]

protozoa   (proh-tuh-ZOH-uh) [plural]

protozoan   (proh-tuh-ZOH-uhn) [singular]

www.ingramcontent.com/pod-product-compliance
Lightning Source LLC
Chambersburg PA
CBHW040151200326
41520CB00028B/7572